DODGE TRUCKS
1948 THROUGH 1960
PHOTO ARCHIVE

DODGE TRUCKS 1948 THROUGH 1960

PHOTO ARCHIVE

Photographs from the
Iconografix Collection of Automotive Images

Edited with introduction by
P.A. Letourneau

Iconografix
Photo Archive Series

Iconografix
PO Box 609
Osceola, Wisconsin 54020 USA

Library of Congress Card Number 95-77489

ISBN 1-882256-37-9

95 96 97 98 99 00 5 4 3 2 1

Book and cover design by Lou Gordon, Osceola, Wisconsin

Printed in the United States of America

Book trade distribution by Voyageur Press, Inc. (800) 888-9653

PREFACE

The histories of machines and mechanical gadgets are contained in the books, journals, correspondence and personal papers stored in libraries and archives throughout the world. Written in tens of languages, covering thousands of subjects, the stories are recorded in millions of words.

Words are powerful. Yet, the impact of a single image, a photograph or an illustration, often relates more than dozens of pages of text. Fortunately, many of the libraries and archives that house the words also preserve the images.

In the *Photo Archive Series*, Iconografix reproduces photographs and illustrations selected from public and private collections. The images are chosen to tell a story-to capture the character of their subject. Reproduced as found, they are accompanied by the captions made available by the archive.

The Iconografix *Photo Archive Series* is dedicated to young and old alike, the enthusiast, the collector and anyone who, like us, is fascinated by "things" mechanical.

ACKNOWLEDGMENTS

Certain photographs were provided by Howard L. Applegate, distinguished dealer in original automotive literature. The captions and photographs were reviewed by Don Bunn, venerable Dodge truck historian. We appreciate their contributions.

A portion of the 1954 Dodge light-duty truck lineup: half-ton pickup, one-ton stake, and half-ton Town Panel.

INTRODUCTION

Although truck production for the military continued throughout World War II, Dodge did not build trucks for the general civilian population between April 1942 and April 1945. When unbridled distribution resumed, the trucks were virtually unchanged from those of 1942. Industry sales were strong, nonetheless, as demand for new vehicles was at an all-time high. (Dodge was not the only manufacturer to enter the post-war era with pre-war models. All manufacturers were in the same position, having committed few resources to new model design during the war.)

A full range of new Dodge trucks, from half-ton pickups to high tonnage models, arrived for the 1948 model year. The company touted the new trucks as "brilliant" and "strikingly new." Although there were a number of mechanical improvements that affected their operation and performance, the trucks were most distinguished by their new cabs. Increased glass area, with higher and wider windshields and windows, improved ventilation, all-steel construction, and more comfortable seating were hallmarks of the new *Pilot-House* cab.

A new series of Dodge trucks appeared in 1954, with the introduction of the V-8 pickup. The C-Series pickups offered an optional V-8 engine and a re-styled, lowered cab and low-sided box, and single-piece windshield. The Town Panel and Town Wagon were also new to the C-Series, the former debuted in 1954 and the latter in 1955. In mid-year 1955, a number of new changes were introduced, that included "wrap around" windshields and PowerFlite automatic transmissions. This new series remained virtually unchanged for the 1956 model year.

In 1957, a facelift resulted in the hooded headlight that distinguished all Dodge trucks of that year. Also introduced was the celebrated D-100 *Sweptside*, arguably the most beautiful pickup of the 1950s. The next revolution in Dodge pickup styling followed in 1959, with the introduction of the *Sweptline*. Its full-width box, all-steel floor, and extra-wide tailgate made the *Sweptline's* clean design among the most practical of its era.

In 1960, Dodge introduced the modern looking *Cab-Forward* to its medium and heavy-duty lines. This compact design combined the benefits of conventional and cab-over-engine designs while eliminating the drawbacks. It offered short wheelbase maneuverability and full-size cab comfort. Its exclusive *Servi-Swing* fenders provided optimum access to either side of the engine for service and maintenance.

It was not possible to include a photograph of every model Dodge truck built in this post-World War II era. A sincere effort was made to include a variety of chassis and body styles, as a means to demonstrate why the Dodge trucks of this period are among today's most popular collector vehicles.

1948 THROUGH 1953
PILOT HOUSE YEARS

1948 Model F-152 one-and-one-half ton, 12-foot stake truck.

1948 one-ton van with deluxe cab.

1948 cab-over-engine tractor trailer used to haul "Mighty Mite", a scale model of a Pacific Intermountain Express rig.

1948 two-ton C.O.E. B-1-KMA chassis with aircraft fueling body.

1949 two-ton B-1-KA with van body.

1949 two-and-one-half-ton B-1-TA with van body.

Two 1950 one-ton Power Wagons and a 1950 Ford F-1 pickup.

1951 Model B-3-D-126 one-ton tow truck.

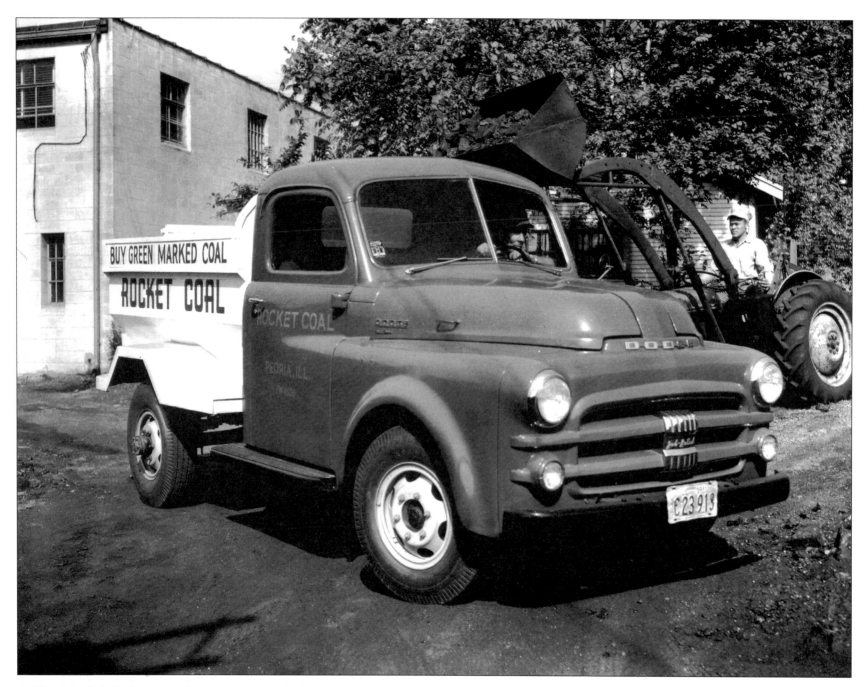

1951 Model B-3-D-116 one-ton coal delivery truck.

1951 two-ton B-3-HH-152 coal delivery truck.

1951 two-and-one-half-ton B-3-KA-152 coal delivery truck.

Two views of a 1951 two-ton Model B-3-HHA-128 with nine-foot stake body.

School bus body on a 1951 Model F5-152 school bus chassis.

1952 Model B-3-HHA-152 chassis cab with twelve-foot platform.

1952 tractor trailer units, center and right, and a 1950 model on left.

1949 C.O.E. wrecker and a 1953 Dodge tractor-trailer rig used to carry the Jimmy Lynch "Death Dodgers" 1953 Dodge convertible.

Two views of a 1953 B-4 series half-ton pickup with Truck-O-Matic four-speed semi-automatic transmission.

Dodge dealership truck lot, *circa 1953.*

1954 THROUGH 1960
V-8 AND POWER GIANT YEARS

1954 Dodge half-ton pickup with optional V-8.

1954 Job-Rated half-ton Town Panel with 145-horsepower Power-Dome V-8 engine.

1954 De Luxe half-ton Town Panel.

1954 three-and-one-half-ton 132-inch wheelbase C-1-V8-132 chassis cab with fifth wheel and tandem trailer.

1954 or *first-series* 1955 one-ton nine-foot stake truck.

Second-series 1955 Custom Regal half-ton pickup.

Rear and side views of a *second-series* 1955 Regal half-ton pickup.

Second-series 1955 half-ton Town Wagon.

Second-series 1955 Custom Regal half-ton Town Panel.

Second-series 1955 Custom Regal half-ton pickup.

1956 Custom Regal half-ton Town Panel.

1956 Model H two-ton chassis cab.

1956 Model G one-and-one-half-ton twelve-foot stake truck.

1956 Model KM two-and-one-half-ton C.O.E chassis cab.

1956 Model YX six-wheeled four-ton cement mixer truck.

1957 D-100 half-ton pickup with Power-Dome V-8.

1957 D-100 with optional side-mounted spare tire.

1957 D-100 Town Panel.

1957 D-100 Town Panel.

1957 D-300 one-ton *dually* pickup.

1957 P-300 forward control delivery van.

1957 D-600 two-and-one-half-ton twelve-foot stake truck.

1957 D-500 two-ton van.

School bus built on a 1957 S-700 236-inch wheelbase chassis.

1957 C-700 three-ton C.O.E. tractor.

1957 T-900 three-ton tandem dumper.

American LaFrance fire truck built on a 1957 D-900 chassis.

A standard D-800 or standard D-900 chassis.

1958 Sweptside D-100 pickup.

A favorite Dodge export, a 1958 Fargo 100 half-ton pickup.

1958 D600 two-and-one-half-ton with grain body.

1958 Model T-700 tandem dumper.

1959 Sweptline D-100 half-ton pickup.

1959 Sweptline D-100 half-ton pickup.

1959 Sweptside D-100 half-ton pickup.

1959 D-100 half-ton Town Panel.

1959 D-100 half-ton Town Panel.

1959 Fargo 100 half-ton Town Panel.

1959 Dodge W-100 half-ton Power Wagon Town Wagon.

1959 W-300 one-ton Power Wagon stake truck.

American LaFrance fire truck body on a W-300 one-ton Power Waron chassis cab.

Two 1959 P-300 one-ton forward-control step vans with Boyertown bodies.

1959 D-300 one-ton platform.

1959 D-400 with horse body.

Hard-working 1959 one-and-one-half-ton D-400 stake trucks.

1959 D-400 van.

1959 D-500 two-ton fuel oil delivery truck.

1959 D-600 two-and-one-half-ton chassis cab with van body.

1959 D-700 three-ton dump truck with Power Giant 218-horsepower 354-cubic inch Hemi V-8.

1959 D-700 dumper with a maximum
GVW rating of 25,000 pounds.

1959 Dodge D-800 three-and-one-half-ton platform hauling cement blocks.

1959 Dodge D-800.

1959 Dodge T-900 four-ton tractor and tanker.

1959 Dodge T-900 tandem.

1959 Dodge T-900 tractor.

1960 Dodge D-100 half-ton Sweptline pickup.

1960 D-100 Town Wagon.

1960 D-200 three-quarter-ton Tradesman.

1960 D-300 one-ton chassis cab with wrecker body.

1960 Dodge D-500 van.

1960 Dodge D-400 refrigerated van.

1960 Dodge D-500 stake truck.

Rendering of a 1960 Dodge D-700 fire truck.

1960 C-SERIES
CAB FORWARD DESIGN

Rendering of a 1960 CN-1000 tractor hauling a missile.

1960 C-500 tractor.

1960 C-500 straight-truck
with stake body.

1960 D-500 van and 1960 C-600 van.

1960 C-700 tanker.

1960 C-700 tanker and 1960 D-100 Sweptline half-ton pickup.

1960 C-800 tractor trailer and 1957 D-800 tractor.

1960 CN-800 tractor trailer.

1960 CT-900 tractor.

1960 CT-900 tractor and 1959 D-100 half-ton pickup.

1960 CNT-900 tractor trailer.

As is evident on this 1960 CNT-1000 tractor, accessibility was the hallmark of the *Cab-Forward* C-Series and its full-opening hood and *Servi-Swing fender*.

The Iconografix Photo Archive Series includes:

TRACTORS AND CONSTRUCTION EQUIPMENT

CASE TRACTORS 1912-1959 *Photo Archive*	ISBN 1-882256-32-8
CATERPILLAR MILITARY TRACTORS VOLUME 1 *Photo Archive*	ISBN 1-882256-16-6
CATERPILLAR MILITARY TRACTORS VOLUME 2 *Photo Archive*	ISBN 1-882256-17-4
CATERPILLAR SIXTY *Photo Archive*	ISBN 1-882256-05-0
CATERPILLAR THIRTY *Photo Archive*	ISBN 1-882256-04-2
FARMALL F-SERIES *Photo Archive*	ISBN 1-882256-02-6
FARMALL MODEL H *Photo Archive*	ISBN 1-882256-03-4
FARMALL MODEL M *Photo Archive*	ISBN 1-882256-15-8
FARMALL REGULAR *Photo Archive*	ISBN 1-882256-14-X
FORDSON 1917-1928 *Photo Archive*	ISBN 1-882256-33-6
HART-PARR *Photo Archive*	ISBN 1-882256-08-5
HOLT TRACTORS *Photo Archive*	ISBN 1-882256-10-7
JOHN DEERE MODEL A *Photo Archive*	ISBN 1-882256-12-3
JOHN DEERE MODEL B *Photo Archive*	ISBN 1-882256-01-8
JOHN DEERE MODEL D *Photo Archive*	ISBN 1-882256-00-X
JOHN DEERE 30 SERIES *Photo Archive*	ISBN 1-882256-13-1
MINNEAPOLIS-MOLINE U-SERIES *Photo Archive*	ISBN 1-882256-07-7
OLIVER TRACTORS *Photo Archive*	ISBN 1-882256-09-3
RUSSELL GRADERS *Photo Archive*	ISBN 1-882256-11-5
TWIN CITY TRACTOR *Photo Archive*	ISBN 1-882256-06-9

TRUCKS

DODGE TRUCKS 1929-1947 *Photo Archive*	ISBN 1-882256-36-0
DODGE TRUCKS 1948-1960 *Photo Archive*	ISBN 1-882256-37-9
MACK MODEL AB *Photo Archive*	ISBN 1-882256-18-2
MACK MODEL B 1953-1966 VOLUME 1 *Photo Archive*	ISBN 1-882256-19-0
MACK MODEL B 1953-1966 VOLUME 2 *Photo Archive*	ISBN 1-882256-34-4
MACK EB, EC, ED, EE, EF, EG & DE 1936-1951 *Photo Archive*	ISBN 1-882256-29-8
MACK EH-EJ-EM-EQ-ER-ES 1936-1950 *Photo Archive*	ISBN 1-882256-39-5
MACK FC, FCSW & NW1936-1947 *Photo Archive*	ISBN 1-882256-28-X
MACK FG-FH-FJ-FK-FN-FP-FT-FW 1937-1950 *Photo Archive*	ISBN 1-882256-35-2
MACK LF-LH-LJ-LM-LT 1940-1956 *Photo Archive*	ISBN 1-882256-38-7
STUDEBAKER TRUCKS 1928-1940 *Photo Archive*	ISBN 1-882256-40-9
STUDEBAKER TRUCKS 1941-1964 *Photo Archive*	ISBN 1-882256-41-7

AUTOMOTIVE

AMERICAN SERVICE STATIONS 1935-1943 *Photo Archive*	ISBN 1-882256-27-1
IMPERIAL 1955-1963 *Photo Archive*	ISBN 1-882256-22-0
IMPERIAL 1964-1968 *Photo Archive*	ISBN 1-882256-23-9
LE MANS 1950: THE BRIGGS CUNNINGHAM CAMPAIGN *Photo Archive*	ISBN 1-882256-21-2
SEBRING 12-HOUR RACE 1970 *Photo Archive*	ISBN 1-882256-20-4
STUDEBAKER 1933-1942 *Photo Archive*	ISBN 1-882256-24-7
STUDEBAKER 1946-1958 *Photo Archive*	ISBN 1-882256-25-5

The Iconografix Photo Archive Series is available from direct mail specialty book dealers and bookstores throughout the world, or can be ordered from the publisher. For additional information or to add your name to our mailing list contact:

Iconografix
PO Box 609
Osceola, Wisconsin 54020 USA

Telephone: (715) 294-2792
(800) 289-3504 (USA and Canada)
Fax: (715) 294-3414

US book trade distribution by Voyageur Press, Inc. (800) 888-9653

MORE GREAT BOOKS FROM ICONOGRAFIX

IMPERIAL 1955-1963 *Photo Archive*
ISBN 1-882256-22-0

IMPERIAL 1964-1968 *Photo Archive*
ISBN 1-882256-23-9

STUDEBAKER 1933-1942
Photo Archive ISBN 1-882256-24-7

STUDEBAKER 1946-1958
Photo Archive ISBN 1-882256-25-5

DODGE TRUCKS 1948-1960
Photo Archive ISBN 1-882256-37-9

STUDEBAKER TRUCKS 1928-1940
Photo Archive ISBN 1-882256-40-9

STUDEBAKER TRUCKS 1941-1964
Photo Archive ISBN 1-882256-41-7

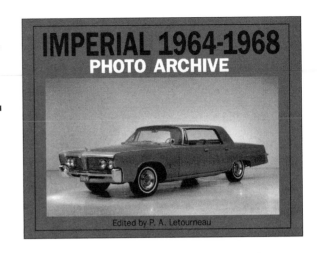

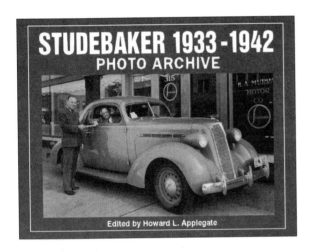

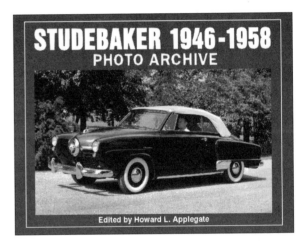

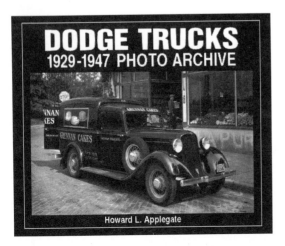

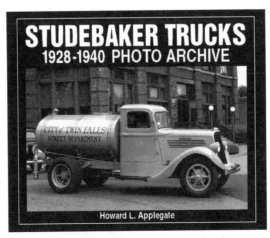

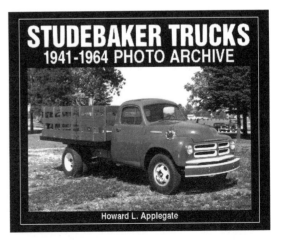

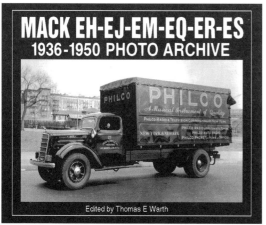
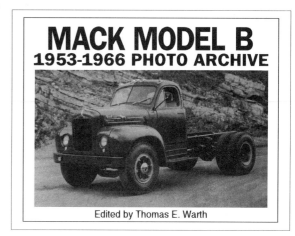
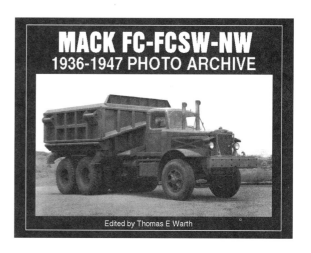
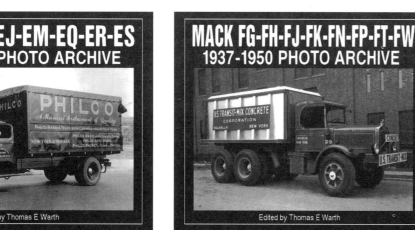
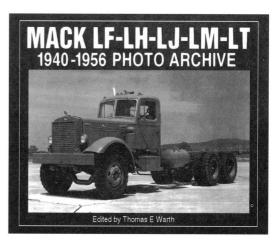